KT-583-333

21

Alcohol

WITHDRAWN

Emma Haughton

THE LEARNING CENTRE
HAMMERSMITH AND WEST
LONDON COLLEGE
GLIDDON ROAD
LONDON W14 9BL

0181 741 1688

Hammersmith and West London College

304679

Talking Points series
Alcohol
Animal Rights
Charities – Do They Work?
Divorce
Genocide
Homelessness
Mental Illness
Slavery Today

Editor: Jonathan Ingoldby
Series editor: Alex Woolf
Designer: Simon Borrough
Consultant: Fiona Feehan

First published in 1998 by
Wayland Publishers Ltd, 61 Western Road,
Hove, East Sussex, BN3 1JD, England.

© Copyright 1998 Wayland Publishers Ltd

Find Wayland on the Internet at
http://www.wayland.co.uk

**British Library Cataloguing in
Publication Data**
Haughton, Emma
    Alcohol. - (Talking points)
    1.Alcohol - Juvenile literature
    2.Alcoholism - Juvenile literature
    3.Alcohol – Social aspects – Juvenile
    literature
    1.Title
    362.2'92

ISBN 0 7502 2182 8

Printed and bound in Italy by G.Canale &
C.S.p.A., Turin

Acknowledgements
With thanks to my husband Jonathan Rees for
all his diligent research.

Picture acknowledgements
Allsport 29 (Mark Thompson), 57 (Phil Cole);
Sally and Richard Greenhill cover, 4, 11, 31,
32, 33, 36, 37 (Kate Mayers), 55b; Angela
Hampton Family Life Pictures 12, 13, 15, 18,
19, 20, 21b, 34, 44t, 54, 55t; Robert Harding
Photo Library 45; Michael Holford 5t, 6;
Impact Photos Ltd 7(Geraint Lewis), 8 (John
Norman), 9 (Andy Johnstone), 10 (Andy
Johnstone), 22 (Eliza Armstrong), 23 (David
Gallant), 24(Bruce Stephens), 25t (Caroline
Penn), 25b (Paul Forster), 26 (Alexis
Wallerstein), 27 (Simon Shepheard), 28
(Michael Gover), 30 (Andy Johnstone), 35
(Mark Henley), 40 (Martin Black), 44b (Simon
Shepheard), 46 (Petteri Kokkonen), 47 (Rachel
Morgan), 48 (Simon Shepheard), 49 (Caroline
Penn), 51 (John Cole), 52 (Simon Shepheard),
53 (Homer Sykes), 56 (Stuart Clarke);
Network Photographers 16 (M. Peterson), 41
(M. Peterson), 59 (M. Peterson); Chris Schwarz
14, 21t, 58; Science Photo Library 42 (Peter
Menzel); Shout 43, 50; Topham Picturepoint
38; Wayland Picture Library 5.

HAMMERSMITH AND WEST
LONDON COLLEGE
LEARNING CENTRE

20 JUN 2000

SHJ 1793 £10.99
304679
362.292
Reference

Contents

What is alcohol?

Alcohol is made by the process of fermentation, in which sugar is converted into ethanol. Ethanol alcohol has a powerful effect on the human body and mind. For centuries humans have produced alcoholic drinks not only because they taste good, but because they can change the way we think and feel.

You can make alcohol from almost any edible substance. Different cultures have used different foodstuffs as the base from which to brew alcohol. The Saxons, for instance, made mead from honey, while the Japanese still make wine from rice, called *sake*. Stronger alcohols, such as whisky and gin, are formed by distillation, in which alcohol is heated to concentrate the levels of ethanol. This process has been used from as early as the tenth century.

Wine grapes like these have been used to make wine for thousands of years. Today they are grown all over the world, resulting in a huge variety of wines from many different countries.

Since ancient times, civilizations across the world have discovered the secret of producing alcohol and incorporated it into their diet and culture. The earliest evidence of wine – alcohol fermented from grapes – comes from the Middle East almost 6,000 years ago.

Wine was a very important part of Roman culture: for the first two centuries BC, wine was exported from Italy to all parts of the Roman Empire, in exchange for the slaves required to cultivate the grapes and make the wine on the large villa estates. In Rome, sweetened wine called *mulsum* was freely distributed at games and circuses to encourage political support. After a while, the demand for wine became so high in Rome that it had to be imported from Spain and France.

This Roman mosaic from the fourth century AD shows grapes being harvested and trodden in readiness to make wine.

The Romans did not just drink their wine. Recognizing its antiseptic properties, they used it to bathe the wounds of gladiators injured in combat.

The Romans perfected the technique of storing wine in *amphorae* (clay jars) with cork stoppers. After the fall of the Roman Empire this method of preserving wine was not rediscovered until the seventeenth century.

Wine in the Roman Empire

Roman wines were as varied and sophisticated as many we drink today. There were also many vintages. The most praised was Falernian, made from the Aminean grape and grown on the slopes of Mount Falernus in Italy. Julius Caesar's favourite, however, was said to be Mamertine from Messina, which he served at public banquets.

Beer, which is made using grain, also has a long tradition; according to Egyptian myth, Osiris, the god of agriculture, taught humankind the art of brewing beer. For many centuries in Europe, ales were drunk in place of water, because there was less risk of the drinker picking up a disease. In medieval times weak beer was even given to children as a normal part of their diet.

Talking point

'Drunkenness is nothing but voluntary madness.'

Seneca, Roman philosopher, 4 BC–AD 65

For as long as people have produced alcohol, they have been aware of the negative effects it can have on the human mind and body. If drunkenness is a state of madness, why do people so often enter it willingly?

These funerary models show that beer was being made in Egypt as early as 2250 BC.

Some people would argue that Christianity and alcohol are incompatible, but wine is actually an important part of worship.

Alcohol and religion

The central role that alcohol, particularly wine, played in many early cultures can be seen in religious texts. In Christianity, the Bible has opposing views on drinking, but the importance of wine to the Jews and later the Christians is evident throughout.

Christians believe that Jesus turned water into wine for one celebration; today priests still use wine to represent the blood of Christ in the Holy Communion service or Eucharist. Although monks are often seen as turning away from the material pleasures of the world, some of the world's finest beers are still brewed today in Trappist monasteries.

The conflicting views of religion

'Believers, wine and games of chance, idols and divining arrows are abominations, devised by Satan. Avoid them, so that you may prosper.'

Qur'an, verse 90

'A man hath no better thing under the sun, than to eat, and to drink, and to be merry.'

Ecclesiastes, 8:15

'Woe unto them that rise up early in the morning, that they may follow strong drink; that continue until night, till wine inflame them.'

Isaiah, 5:11

Some religions have always opposed the use of alcohol in society. Muslims are forbidden to drink it by their holy book, the Qur'an. In some countries, such as Saudi Arabia, this ban is taken so seriously that drinking alcohol is illegal. Hindus, Mormons and devout Buddhists also reject alcohol for religious reasons.

In Arab culture, socializing does not revolve around alcohol, but around drinks like coffee and tea.

Types of alcohol

The types and prices of alcoholic drinks vary enormously around the world. It is possible to brew your own beer or ferment your own wine very cheaply, for just the cost of the ingredients. On the other hand you can easily spend large sums of money on a rare and prized vintage of wine. Generally, however, the price depends on how much alcohol a drink contains: the more alcohol it has in it, the more expensive it is.

There are many different types of alcohol available today: a wide variety of wines, from pale fizzy champagnes to heavy red clarets and burgundies; hundreds of beers and ales, from light German lagers to dark Irish stout; and distilled alcohols (called spirits) which include whisky, brandy, gin and vodka.

A huge range of alcoholic drinks are available, and people's tastes vary widely, as can be seen in this busy bar in Prague in the Czech Republic.

Some spirits are blended with other alcohols to produce different drinks: sherry, for example, is a mix of brandy and wine. Alternatively, spirits can be flavoured to produce liqueurs such as orange-flavoured Cointreau and aniseed-flavoured Pernod, or mixed with other ingredients such as cream, coffee or egg to produce popular drinks such as Baileys or Advocaat.

More recently, brewers have developed a range of new alcohols, known as 'alcopops', based on soft drinks such as cola and lemonade. With a sweet flavour much like their non-alcoholic counterparts, these drinks are heavily aimed at younger people, who have not yet developed a taste for conventional alcohols such as wine, beer and spirits.

Worldwide consumption of alcohol

In some countries, levels of drinking are rising rapidly. More people in Eastern Europe are now buying alcohol, where sales have reached over $20 billion a year compared with $15 billion five years ago. In Russia alone, official statistics show that the amount of alcohol consumed per person has gone up 600 per cent in the last ten years. In other parts of the world, however, alcohol consumption is still very low. The countries that drink least are those where alcohol is either illegal or heavily discouraged, principally those in the Middle East, or where it is highly taxed, such as in Norway and Sweden.

Who consumes the most?
- Europe, the USA, Canada and Australia have the highest alcohol consumption overall.
- Germany consumes the most beer: the average person consumes over 130 litres a year.
- France, Italy, Spain and Portugal consume the most wine, although the amount has fallen over the last 50 years due to rises in the legal drinking age and people's concern for their health.

Beer plays an important part in German culture, and Germany produces both pale lagers and dark, strong, slightly sweet beers.

An important part of our lives?

In many parts of the world, alcohol is perhaps more important than ever. In restaurants wine is seen as an essential accompaniment to a good meal, while for the many thousands of bars and pubs around the world, alcohol is the mainstay of their business. Indeed, going out to a bar or pub is still one of the most popular forms of entertainment in many countries, and getting drunk at Christmas and New Year parties is something that many people see as entirely normal, and often as a sign that they have had a good time.

Even in the home, drink is seen as an important part of social and family life. Many people, for instance, feel that opening a bottle of champagne or sherry is an essential part of celebrating or marking any great success or life event, such as a marriage, a birth, or even a funeral.

A bottle of champagne is often seen as an essential part of a family celebration, such as this eighteenth birthday party.

The effects on your body

Good for your health?

'Tobacco in every circumstance is bad for you and those around you. There is considerable evidence that alcohol consumed in moderate quantities is not harmful and may even be good for your health.'

Barbara Nolan, spokesperson for EU Health Commissioner Padraig Flynn, 15 January 1998

Alcohol is a powerful drug. Although it comes in many different forms and strengths, all alcohol has basically the same effects on the human body and mind. Some of these can be quite pleasant, and are often among the reasons people choose to drink. Mild intoxication, known as being 'tipsy', can make you feel more relaxed, sociable, happy and confident. That's one reason why people drink more at social occasions – a few drinks can make it easier for them to 'break the ice' and really enjoy themselves.

A little of what you fancy does you good?

Drinking alcohol is not always bad for you. Many countries consider drinking under a certain level acceptable. Indeed, the UK government recently raised the recommended guidelines on safe drinking levels to between three and four units a day for men, and two and three units a day for women. The government said there was no evidence that drinking up to these levels would put someone's health at risk.

Some studies have even shown benefits from moderate drinking. An Australian study of over 16,000 people found that the risk of heart attack is lowest for men who have between four and five units of alcohol a day, five or six days a week. One study even showed that heavy drinking protected men against going bald.

Moderate drinking is regarded as beneficial in many countries, and the initial effects of alcohol can be very pleasant.

Alcohol units

Drinking sensibly means being aware of roughly how much alcohol there is in a drink. You need to bear in mind, however, that beers and wines can vary a lot in strength. Extra-strength lager, for instance, contains almost three times as much alcohol as ordinary lager.

1 unit = ½ **pint of beer**
= **1 small glass of wine**
= **single measure of spirits,
e.g. whisky, gin, vodka**

The alcohol content of a drink is referred to on the bottle or the can as a percentage. On a can of beer, for instance, it may say 3.5 per cent, indicating that it is roughly ordinary-strength beer; figures much higher than this mean that it will take a lot less to get you drunk. The strongest drinks are spirits, which can have an alcohol content of up to 40 per cent i.e. nearly half the liquid is pure alcohol.

On average it takes an hour for your body to get rid of the alcohol in one unit.

Bad reactions

Unfortunately, people can also have very negative reactions to alcohol. The more you drink, the more unpleasant the experience is likely to become.

Alcohol is a depressant, and it can impair your judgement, slow your reactions and make it more difficult to perform physical tasks. You can lose your co-ordination, and feel less pain if injured. It is hardly surprising that when we think of someone who is drunk, we picture them staggering around, unable to walk in a straight line. When you are drunk, your speech is likely to slur and your memory may be temporarily affected.

In the later stages of drunkenness people often vomit, and can find it difficult to do even the most basic things, like recognize people they know or remember their own name and address.

Excessive alcohol consumption can make you almost helpless, and unable to move properly, think clearly or remember the simplest things.

Case study

In the accident and emergency department of Liverpool Children's Hospital, Chloe, 14, sits shivering in a thin blouse and jeans. She is very drunk. A policeman found her slumped on the pavement, nearly unconscious, her clothes splattered with vomit and drink. She shouts and slurs her words as she is moved to the resuscitation department.

The doctor asks her who she is and how much she has drunk, but Chloe cannot remember much about what she has been doing. She just knows she went out with her friends and they were drinking cider. Then they persuaded someone to buy vodka for them from a shop.

A woman sits down beside her with tears in her eyes, but Chloe does not recognize her mum. When a nurse inserts a drip into her arm, Chloe does not even look up. Her mum is worried about whether Chloe will come

In the UK, 1,000 children a year are admitted to the emergency units of hospitals after drinking alcohol.

to any permanent harm. The nurse reassures her that on this occasion Chloe will get away with just a bad hangover the next day, but warns that if she drinks this heavily again there is a chance that it might kill her.

Talking point

'I like liquor – its taste and its effects – and that is just the reason why I never drink it.'

Thomas 'Stonewall' Jackson, a general in the US Civil War

Some people know that it is easy to have too much of a good thing, and would rather abstain from alcohol altogether. Do you think alcohol should be banned for health reasons?

If you drink too much too quickly, alcohol can actually kill you. Alcohol poisoning occurs when the brain becomes so affected that the heart or breathing stops. You can also choke to death if you vomit, or suffer heart failure as a result of being suddenly sick. One and a half bottles of spirits are enough to kill the average person, if drunk quickly; if you are small, female or young, you will be in danger from a lot less.

Blood alcohol

The amount of alcohol you have drunk can be measured quite accurately in your blood. The amount in your bloodstream is related to how drunk you feel and the symptoms you experience, although it varies according to how big you are, your sex and other factors, such as how much you have eaten.

The bigger you are, the more alcohol you can tolerate in your bloodstream before feeling the effects. Women will experience more symptoms at a lower blood level because they tend to be smaller and have a lower proportion of water in their bodies to dilute the alcohol. Having a full stomach will slow down the rate at which you absorb alcohol from your stomach into your blood. The effects also depend on how used you are to drinking.

The speed at which you drink also influences the effects on your body. Drinking competitions like this one are probably one of the worst ways to drink alcohol.

Blood alcohol levels

This table shows levels of blood alcohol and their effects on an adult male of average build (a third less blood alcohol would have the same effects on an adult female of average build).

Blood alcohol (mg/100 ml)	Beer drunk (pints) (1 pint = 0.57 litres)	Effects
20	$^1/_2$–$^3/_4$	Feel good and relaxed. Little effect on performance.
50	$1^1/_2$	Feelings of sociability and friendliness. May slightly affect driving performance.
60	2	Judgement and decision-making start to become impaired. Driving becomes reckless.
80	$2^1/_2$–3	Loss of co-ordination. Completely unsafe to drive.
100	3	May become sleepy.
160	5	May become aggressive, behaviour unmanageable. Subsequent memory loss is likely.
300	10	May lose control of bodily functions, e.g. continence. Danger of coma.
500	18	Likely to die without urgent medical help.

Alcohol and your emotions

Many of the problems arising from alcohol result from its effect on your emotions and feelings. Once you drink enough to lose some of your inhibitions, you are likely to keep on drinking and care less and less about the consequences. Alcohol can also depress your mood, making you feel unusually sad or particularly aggressive. Many suicide attempts and assaults on others occur when someone has been drinking. You are more likely to do or say things that you will later regret, and to take risks that you would ordinarily avoid.

Alcohol can temporarily numb bad feelings, but often you end up feeling much worse.

The day after

If all that were not bad enough, there is the next day to face. Although hangovers tend to get worse as you get older, the young are not spared if they drink too much. People may treat hangovers as something of a joke, but having one can be awful. The pounding headache, queasy stomach, bad mood and feelings of exhaustion are all signs of the extent to which the alcohol has poisoned your body. Alcohol causes dehydration (loss of water), uses up vitamins, and puts excessive strain on your body. It can take all day to recover.

Having a bad hangover is no joke.

Longer-term effects

In the longer term, the effects of excessive alcohol consumption are far more serious. Drinking heavily over a number of years puts a great strain on the liver, the organ in your body which breaks down and eliminates toxins.

Regular alcohol misusers can develop cirrhosis, where the liver becomes so scarred that it can no longer function, and without a liver transplant people often die. People in countries where a large amount of alcohol is consumed, such as France, Germany and Italy, have the highest death rates from liver disease.

Alcohol-related deaths

- In Finland, between 1985 and 1990, alcohol-related deaths constituted 11 per cent of all deaths among men aged 20 and over.
- In 1993, 24,000 people died in the USA from liver disease and cirrhosis; some 3,000 died in the UK.
- Life expectancy for men in Russia has fallen from 73 years to 58, largely due to increased excessive drinking.

Many men accept a 'beer gut' as a natural, if unattractive, part of middle age. However, as the name suggests, the true cause is usually too much drink.

Heavy drinking can have other effects on health, including liver cancer, hepatitis, heart disease, stomach problems such as bleeding and ulcers, and kidney failure. It can also cause sexual difficulties such as impotence in men, and has been identified as a significant risk factor in breast cancer in women. Alcohol is also very fattening; two pints of beer contain more calories than the average chocolate bar. In the USA, alcohol often carries a health warning in the same way as cigarettes in other countries, and some people argue that this approach should be extended across the world.

Drinking during pregnancy

Drinking heavily during pregnancy can cause serious damage to the unborn child. At its worst, drink can cause foetal alcohol syndrome (FAS), where the baby is born severely mentally and physically handicapped. Each year nearly 10,000 newborn babies are diagnosed in the USA as suffering from FAS; in fact, alcohol is said to be the leading environmental cause of babies being born physically or mentally handicapped in the Western world. In some American states women have actually been prosecuted for continuing to drink after being warned that it would damage their babies.

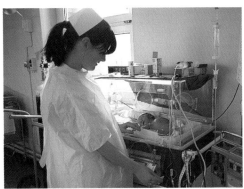

For most people it is their decision whether or not to drink, but what choice did this baby have about its mother's heavy drinking?

No one has yet proved that any level of alcohol is safe in pregnancy, so many women do not drink at all while expecting a child. A lot of women worry that a heavy bout of drinking before they realized they were pregnant may have damaged their baby, but fortunately this is rarely the case. Some people believe that women are made to worry unnecessarily about drinking in pregnancy, and point out that many pregnant women do drink in moderation with no apparent ill effects for their babies at all.

Despite the possible risks, some mothers feel it is safe to drink in moderation during pregnancy.

A bad habit

One of the worst consequences of long-term heavy drinking is that people can become dependent on alcohol. People who drink too much are known as alcoholics. No one is quite sure why some people become alcoholics. It may be that they have a genetic tendency to dependence on alcohol, psychological problems, or that they are constantly in contact with alcohol in social situations. Experts have now come to see alcoholism as a disease in its own right.

Alcoholism usually develops over a number of years: early symptoms include always needing to have drink available and being unusually able to tolerate its effects. As the disease progresses, the people affected lose control over their drinking, which affects their work, social and family life. They may experience severe withdrawal effects if they stop drinking for any reason. The symptoms of withdrawal from alcohol can actually be severe enough to kill.

Alcoholism can cause great pain for sufferers and their families. Despite evidence to the contrary, there remains a tendency to place all the blame for alcoholism on the alcoholic, who is often seen as being able to stop if they 'really wanted to'.

Alcoholism can lead to family break-up and bankruptcy. Many victims end up on the streets.

The economics of alcohol

Harvesting grapes provides plentiful seasonal work in rural areas.

No livelihood

'Alcohol is simply the most important thing in my life. Without it I wouldn't have a livelihood.'

Tony, English pub owner

Alcohol is big business. Around the world millions of people owe their livelihoods to it. For many countries, exporting their national and regional drinks brings in a great deal of revenue. It also provides much-needed work for people, especially in rural areas where jobs might otherwise be scarce.

Russia sells vodka, its traditional national spirit, all over the world. Scotland is famous for its different whiskies. In France, Spain and Italy, the production of hundreds of types of regional wines provides employment for many thousands of rural farmers and vineyard workers.

Case study

Philippe is 17 years old. Last year he left school to work with his father in the family business producing cognac, the famous brandy made in the Charente region of France. His father's business is one of over 200 brandy distilleries in the region, providing one of the main sources of work in a rural area with otherwise high rates of unemployment.

It will take many years for Philippe to master the delicate art of blending the brandy to get just the right quality and taste. He will also learn how to market and export the finished product across France and to other countries – over 80 per cent of the brandy Philippe's father produces is exported abroad. But Philippe is keen to learn. He is grateful to have interesting work that is both secure and profitable. Many of Philippe's old school friends have already had to leave the area to find a job in the larger cities.

Unlike many of his friends, Philippe can remain in the Charente region because of the work provided by his father's business.

It is not just growing the raw materials and making drink that provides work. Many more jobs are involved in distributing alcohol at home and abroad, and selling it to the public through shops, pubs, clubs and bars. Across the Western world, most towns and villages have a place that sells alcohol,

providing jobs for local people. Could pubs, bars and restaurants survive if they did not sell drink?

If it did not sell drink, trade at this popular Greek bar would be almost non-existent.

Alcohol and taxation

Alcohol also makes a lot of money for governments, in the form of taxation. In the UK, for example, where alcohol taxes are high, alcohol sales contribute a significant part of the government's income. While many governments try to discourage people from drinking too much, they would suffer greatly if people stopped altogether. Governments can find it difficult to strike the right balance. They need the money from alcohol taxes, and the jobs that its manufacture, sale and distribution provide.

As the huge size of this distillery in São Paulo, Brazil, shows, alcohol production is a major world industry employing thousands of people in many different jobs.

In the USA, 60,000 jobs were lost in the brewing industry when tax was doubled in 1991. On the other hand, alcohol costs governments a great deal in terms of health costs and traffic accidents. For example, in Spain the misuse of alcohol costs the government the equivalent of over $4 billion each year and in the USA medical care for alcohol-related injury and illness costs at least $28.5 billion a year.

English tourists flock to this unofficial 'cash and carry' in the French port of Calais to stock up on cheap wine and beer.

European competition

Taxing alcohol can cause other problems. Alcohol taxation in countries like France and Spain, for example, is much lower than in many northern European countries. This has created problems with smuggling, where people buy cheap alcohol from countries with a lower tax on alcohol to drink or sell-on in countries where the tax is higher. Since the UK government relaxed restrictions on how much wine and beer people could bring back from abroad, many UK pubs and shops have complained that they are losing vital business because they cannot compete with the cheaper European prices.

Revenue from alcohol

'Why should I feel guilty about drinking, when each drink I buy pays so much tax for this country?'

Chip, US ship welder

- In 1994 in the UK the drinks market was worth £25.8 billion, and raised over £9 billion for the government.
- In the UK, tax accounts for 4 per cent of the price of a bottle of beer, 12 per cent of the price of a bottle of wine and 40 per cent of the price of a bottle of spirits.
- In Finland, alcohol sales generate the equivalent of about £800 million a year in tax for the government.
- The Australian government currently imposes a wholesale sales tax of 40 per cent on alcohol.

'British pubs face a real and growing threat from cheap duty-paid imports of beer from Europe.'

Ian Prosser, chairman, Brewers and Licensed Retailers Association

The UK has roughly one pub per 550 adults over 16, and around 1 million people are employed in alcohol-related jobs. Pubs are seen to be under threat from cheap duty-paid alcohol imported from Europe.

Advertising

Although many people argue that the misuse of alcohol can be just as damaging as cigarettes, generally restrictions on advertising drink are fewer. Tobacco advertising at European sporting events has been outlawed for eight years, but the European Union (EU) has refused to place restrictions on alcohol adverts.

Alcohol also brings in a lot of money for advertisers, who are employed by alcohol companies to promote their different brands of drink. Advertisers cash in on the fact that drinking is commonly seen as cool and sophisticated, or at the very least just a bit of harmless fun.

A major influence?

● In 1996 advertisers spent around $900 million promoting alcohol in the USA.

● In the USA, 56 per cent of students in grades 5–12 say that alcohol advertising encourages them to drink.

In Tokyo, as elsewhere, alcohol advertisers like to associate their product with youth and popularity.

Talking point

'The sponsorship doesn't provide any information on the risks associated with these substances. It just looks like a healthy thing to do.'

Judith Strang, San Dieguito Alliance for Drug-Free Youth, USA, 1993

What do you think about alcohol companies sponsoring sport? Does it send the wrong message about alcohol to young people, or is it a harmless way of making sure sports get the money they need to survive?

Sponsorship

The alcohol industry is one of the main sponsors of sport around the world. The money provided by drinks manufacturers keeps many teams, leagues and sports afloat. In return, the alcohol brands benefit from being associated with famous sports people, and the general image of glamour and health that they represent.

In the USA, Anheuser-Busch, which makes the popular Budweiser beer, is one of the major sponsors of American sport. The support of Anheuser-Busch benefits many sports, including women's basketball, triathlon events, motor racing and the American hockey league. Anheuser-Busch was also one of the 12 official sponsors of the 1998 soccer World Cup in France. In the UK, many of the major soccer teams are supported by large brewers like Carlsberg and Guinness. The biggest beneficiary of alcohol sponsorship in the UK is the FA Premiership soccer league, which has a four-year deal worth £36 million with the brewer Carling.

Although alcohol consumption in France is high, it is the only EU nation to have a full ban on alcohol advertising at sporting events.

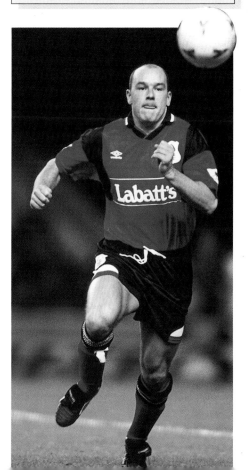

Alcohol manufacturers are keen to associate their product with health, fitness and glamour. Famous sporting personalities provide all three.

Alcohol and young people

Going out for a pint is a normal routine for many young people.

There are a number of reasons why young people start drinking. Alcohol can make you feel more comfortable and relaxed in social situations and when out with friends. A drink can give you the courage to ask out someone you like, or chat to someone you do not know. Often teenagers encourage each other to drink; it is hard to refuse a beer if everyone else in your crowd is having one.

Many teenagers first misuse alcohol simply because they are curious to know what 'being drunk' is like. When everyone around you talks about 'getting plastered' and getting 'out of their heads', it's not surprising

The rise in under-age drinking

A recent survey of UK secondary schools found that there was a 'marked increase' in the number of pupils aged 11 to 15 who drank alcohol every week, with some drinking the equivalent of four pints of beer. Under-age drinking rates are also high in the USA where teenagers typically try their first drink when they are just 13.

that you start to wonder what it feels like and why people seem to enjoy it. A lot of young people believe that drinking somehow makes them more 'grown up' – after all, most adults drink regularly. Some teenagers, particularly those who are under-age, drink because it feels daring and rebellious. In so many situations, such as parties, nightclubs or in a pub or bar, drink is seen as a normal, even essential, part of the experience. No one really stops to question it.

Regular drinkers

● Junior, middle and senior high school students in the USA drink over a billion cans of beer every year.

● In the UK in an average week, 35,000 young people under the age of 16 drink above the safe limits for adults, while 12 per cent of young people aged 9 to 15 are regular drinkers.

The rise in under-age drinking. Some young people are tempted to try alcohol because all their friends are doing the same.

A cause for concern?

Curiosity, rebellion and peer pressure are just some of the reasons why under-age drinking is on the rise in the UK and USA. This is causing a great deal of concern among parents, teachers and medical professionals. But in other European countries such as France and Spain, many young people don't seem so interested in alcohol, despite their countries' high alcohol consumption levels overall.

Proportionally, French and Spanish teenagers actually drink much less than those in the UK and USA. In a study of 7,000 European children aged 11 to 16, 33.3 per cent of the French children and nearly 50 per cent of the Spanish children said they did not drink. This compares with just 14 per cent of British children.

No one knows for certain why this is. It could come down to their upbringing: for example, it is more common for Spanish and French parents to disapprove of their children misusing alcohol. They do not, however, generally disapprove of drinking altogether. They see it as just another part of normal

Teenagers under pressure

'Parents need to be understanding and look at it [alcohol] from the teen's perspective. They need to realize we're facing a whole lot more pressures than they did.'

James Brewster, 17,
Albuquerque, USA

In Europe, young people often drink with their families at mealtimes as part of a cultural tradition.

life, rather than something exciting and special. As a result, French or Spanish children often try alcohol in the company of their family and other adults, in a restaurant or in the home as part of a family meal. This may contribute to their lack of desire to 'go out and get drunk'.

Most parents rarely consider the example they set.

Talking point

'If we are to change the attitudes of young people towards alcohol and the harmful consequences that can result from their drinking then, as adults, we need to start setting a better example.'

Paul McQuail, Alcohol Concern, 15 October 1997

'Parental attitudes towards alcohol appear particularly crucial to boys' consumption, with both overly liberal and too strict regimes encouraging drinking.'

Chris Mihill, *Guardian*, 2 January 1996

What sort of example do the adults around you set? Do you think they are being hypocritical when they warn you about the dangers of drinking too much?

A lucrative market

WARNING
IT IS ILLEGAL TO
PURCHASE ALCOHOL
UNDER THE AGE OF 18

No support

'We have got to stop young people's drinking reaching epidemic levels in some parts of the country. We are clearly getting no support from some of the big players in the business.'

Nigel Griffiths, Labour's consumer affairs spokesman,
Guardian, 4 September 1996

Alcohol manufacturers, who are always seeking ways to sell more of their drinks, are targeting young people more directly in some countries. In the UK, for instance, the alcoholic soft drinks known as 'alcopops' are heavily marketed, often in a way that will appeal to children and young people. Many pub chains now try to attract younger age groups by making their premises more glamorous and attractive places to meet up and socialize.

Alcopops are now widely available and very popular among young people. They are seen by some as being a major contributor to under-age drinking.

Do alcopops encourage under-age drinking?

There has been a lot of concern about alcopops – ordinary soft drinks and pops like coke and lemonade which are made to contain alcohol. Many people believe that the brewers who manufacture alcopops are deliberately encouraging young people to drink.

They argue that whereas traditional drinks like beer, spirits and wine are an 'acquired taste' – that is, they do not taste very pleasant until you get used to them – alcopops are deliberately made to disguise the taste of alcohol and so make it easier to drink. While the makers of alcopops say their drinks are only intended for people aged over 18, one survey found that most of them are drunk by young people aged between 15 and 17. The survey discovered that over half of that age group had drunk at least one alcopop in the last month.

A touch of glamour?

Young people's attraction to alcohol, as with cigarettes, is to a large extent influenced by its image. If you go to see any film at the cinema, or turn on the TV, you are likely to see people drinking. Back in the 1940s and 1950s, many people smoked and it was common to see smoking in films such as *Casablanca,* and on TV. But since many people turned against tobacco for health reasons, cigarette advertising has been banned on TV in both the USA and UK, and smoking is shown much more rarely in films.

In contrast, alcohol has not experienced such a decline. It is still very common to see drink portrayed quite positively in all types of media, including films, books and magazines; and most types of drink can still be advertised on TV. Indeed many teenagers say their favourite adverts are those for beer, which often use wit and humour to promote the product, and specifically aim to appeal to younger people.

Manufacturers will pull out all the stops to make their beer more attractive.

Young and vulnerable

One reason people are so concerned about widespread drinking among young people is because alcohol has particular dangers for children and teenagers. Physically, younger bodies are less able to deal with alcohol, and so its effects are likely to be greater. Although drink can make you feel more relaxed and confident, it tends to intensify any emotion, so can end up making you feel more depressed, violent or aggressive. You may have a few drinks because you want to cheer yourself up, but the more you have, the more you start to dwell on your problems. You can end up feeling much more sorry for yourself than before you started drinking.

35

Case study

'I had my first drink when I was twelve. My friend had taken some beer from home and we went and drank it in the park. I don't remember very much happening. I kept waiting to feel really drunk, but I never did. I guess I didn't have all that much. I felt very disappointed. I really wanted to know what being drunk was like.

Chrissy and her friends always drink at parties. They say that the alcohol makes them feel more relaxed and confident.

I go to a lot of parties round my friends' houses and we always have drink there. I look much older than I really am, so it's not that difficult to get hold of. If we get any trouble, we just ask someone going into the store to get us something, and give them the money. Lots of people are willing to do it, especially if you're really nice to them.

I quite like beer, especially if you put in some lemonade. I prefer sweet drinks, vodka and fruit juice, things like that. I don't get really plastered, well, not most of the time. I just like to feel really relaxed and happy, sort of giggly, you know. You just have a better time if you've had a few drinks, you don't worry so much about how you look or whether boys like you or not. Most of the time we have a real laugh, but sometimes I overdo it and end up being really sick or something. After that I promise myself not to drink again for a long time, but somehow I always do. It's difficult not to, isn't it, when all your friends drink all the time?'

Chrissy, 15

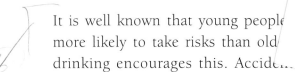

It is well known that young people more likely to take risks than old drinking encourages this. Accide... teenage drivers under the influence of alcohol are a leading cause of teenage death, while many unplanned teenage pregnancies are conceived under the effects of alcohol.

Regular and excessive drinking can also affect your life in other ways, and has been linked to poor school work, social and emotional problems, increased use of other drugs, and even suicide. Adolescence is a difficult time and many teenagers use alcohol as a way of escaping from its pressures and dilemmas.

But sadly, the physical and emotional effects of drinking too much on a regular basis just make it more difficult to cope in the long run. How can you concentrate in class, for instance, if you have a hangover? Drinking heavily if your relationship breaks up will only make you feel more depressed and make it harder to deal with the pain you feel. It is because alcohol intensifies moods that some people try to take their own lives. Their feelings of hopelessness and despair are magnified by the alcohol in their bloodstream.

Drinking alone when already depressed is more likely to magnify your problems, not solve them.

37

strictions on alcohol

For many years there have been laws governing alcohol drinking and purchase in many countries. In the UK, for example, there were drastic changes in the licensing laws at the beginning of the twentieth century. These cut down pub opening hours from nearly 20 hours a day to just over 5, which almost halved the amount of alcohol people drank and reduced alcohol-related deaths by a third. Even today you can only buy alcohol from certain places at certain times.

Prohibition in the USA and the rise of the black market

In the nineteenth century, drinking rates in the USA were high. People drank, on average, 27 litres of pure alcohol a year, compared with around 6.5 litres today. Drunkenness and the resulting crime, poverty and violence were huge social problems.

At midnight on 16 January 1920 the National Prohibition Act came into force, effectively banning the production, sale and consumption of alcohol across the USA. It lasted until 1933, when the need for jobs during the Depression and campaigners for civil rights finally got the Act repealed.

Many people believe, however, that rather than turning the USA into a nation of teetotallers, prohibition had the opposite effect. They claim it promoted disrespect for the law, and encouraged a black market in alcohol

US Government officials destroy a consignment of 749 cases of beer in 1923.

and waves of organized crime. Indeed, it is said that the Mafia got its first foothold in the USA through 'bootlegging', the smuggling of illegal alcohol. However, it is also argued that prohibition was largely effective, reducing drinking levels and alcohol-related crime and injuries.

Many countries also put restrictions on who can manufacture alcohol. In the UK, brewing beer and making home-made wine is legal, but you are not allowed to distil alcohol into stronger drinks such as spirits without a licence. In the USA it is illegal for individuals to make any alcohol at all.

Modern prohibition

The enforcement of alcohol prohibition is not just a thing of the past. On 1 July 1996, prohibition was introduced to the northern state of Haryana in India, in the hope of reducing crime and the harassment of women.

Although supported by many, the ban has also been criticized because some communities have lost so much income as a result of it that they are almost bankrupt. (India is one of the highest consumers of beer and other alcoholic drinks in the world.) Does this show that prohibition is too inflexible and unfair to be a long-term solution to alcohol-related social problems?

Old enough to know better?

Most countries have rules about the age at which you are allowed to start drinking, although just how old you have to be varies according to where you live. These laws can be complex and confusing. In the UK, drinking is completely illegal for children under 5, but older children can drink as long as it is not on licensed premises such as a pub or restaurant. Young people aged 16 to 18 can consume certain drinks, such as wine, in pubs and bars, but only with a meal; it is illegal, however, for anyone under the age of 18 to buy alcohol from an off-licence.

Because he is not yet old enough to drink legally in a bar, this teenager plays it safe and orders a soft drink.

In the USA, the law is strict and simple: young people cannot buy or even possess alcohol at all until they are 21. This can be difficult for people who look younger than they are; they may often be asked in pubs and bars to prove that they are old enough to drink.

Worldwide age limits on drinking

The age limits for alcohol consumption around the world are complex. In some countries you can legally drink some weaker alcohols when you are younger, while drinking stronger drinks is only allowed when you are older. For example, in France you can drink beer at 14, but are not allowed to drink spirits until you are 18. Below is a rough guide to who can drink when.

Country	Minimum age for drinking
USA	21
New Zealand, Japan	20
Canada	18–19
Australia, UK, Brazil	18
France	14–18
The Netherlands	16–18

It's often hard for bar-owners to tell how old their customers really are: do you think all the drinkers in this American bar are over 21?

Minimum drinking-age laws can be a great worry for publicans, bar-owners and shopkeepers, who can get into a lot of trouble if they are found to be selling drink to someone who is under-age. However, such laws are often ignored or flouted, and young people are sold alcohol without being asked their age.

In France the limit on how old you have to be to drink or buy alcohol is lower than in many countries. Some people believe this is the best approach, as it means that young people there do not see drinking as anything special or as proof that they have somehow 'grown up'. Others, however, think that young people need protecting from themselves, and the dangers that alcohol can represent.

Drinking and driving

The tightest restrictions on alcohol relate to drinking and driving, and with good reason. While under-age drinking generally harms no one but the teenagers themselves, drinking while driving can ruin the lives of innocent people.

Every 32 minutes in the USA someone dies in an alcohol-related traffic accident, and every 30 seconds someone is injured. Although sometimes it is the drunk driver who is killed, many of those deaths and injuries involve passengers, other car drivers and pedestrians.

The amount you can drink and still legally drive varies from country to country. In France there are two limits. If you exceed the lower one, it is considered a lesser offence than if you break the higher limit. In Russia, on the other hand, you are not allowed to drive a vehicle if you have had any alcohol at all. In Sweden the limit is so low that even a small drink is likely to make you legally unfit to drive.

Accidental death?

'Have none for the road.'

UK government
anti-drink-driving slogan,
Christmas 1997

● In the UK alcohol is involved in about 15 per cent of traffic deaths.
● In 1994 there were 566 convictions for fatal accidents related to alcohol in France.

You may feel you are fit to drive, but alcohol affects your judgement in many ways.

Talking point

'Any relative of a victim knows the misery caused by drinking and driving.'

Tony Blair,
UK Prime Minister, 1997

If drink-driving causes so much misery, why do people still do it?

Drink-driving restrictions worldwide

Country	Alcohol in mgs per 100 ml of blood
Japan, Russia	0
Sweden, Poland	20
Portugal	40
France, Australia, Belgium, The Netherlands, Greece, Finland, Germany	50
UK	80
USA	80–100

In the UK at the time of writing you will be prosecuted if you are found to be driving with more than 80 milligrams of alcohol per 100 millilitres of blood. In the USA the limit varies from state to state, but is generally between 80–100 milligrams. In some states the limit is lower for young people.

Despite numerous government campaigns in the UK and USA, every year hundreds of children are killed or seriously injured in accidents involving drivers who have consumed more than the legal alcohol limit. The impact on the families is terrible. Not only must they deal with the death of a child, they also have to face the harsh fact that if the driver had not been drinking the accident might have been avoided.

'One for the road' can lead to your last journey.

It is no wonder that such families are often very angry, especially as drunk drivers cannot be charged with murder, only with careless or dangerous driving. Often their court sentence can seem very light considering that their actions have taken someone's life.

43

Emily's parents may never fully recover from the loss of their daughter.

Case study

When their daughter Emily died aged 8 in a UK road traffic accident, her parents, John and Alison, were devastated. The driver hit Emily as she was crossing the road on the way back from school, and left her lying in the road. Luckily a passer-by witnessed the crash, and the driver was eventually caught. But it was too late to help Emily, and she died of her injuries on the way to hospital.

The driver was found to have 75 milligrams of blood alcohol, and he was prosecuted for dangerous driving. His sentence of nine months in prison, a £1,500 fine and two years disqualification from driving, was lighter than he would have received had his blood alcohol been above the legal limit.

John and Alison still think that the driver's deliberate decision to drink and drive directly caused Emily's death, and that he deserved harsher punishment. They do not see why he should be treated so leniently and they think he should have been charged with murder or manslaughter. Alison was on anti-depressants for five years following the crash. John lost his job after suffering from constant ill-health brought on by stress following Emily's death. Emily's brother, Tom, has had difficulties at school ever since she died. The driver left prison after six months.

In 1998, causing death by dangerous driving under the influence of alcohol became a criminal offence carrying a sentence of up to 7 years imprisonment.

Emily is remembered by her friends and family but the driver who caused her death did not even apologize.

How much is too much?

A particular problem is that 'legal' levels of blood alcohol will affect different people in different ways. Although a large man who is used to drinking may be relatively safe to drive at the upper end of the legal limit, a smaller person may be unfit to drive after drinking less. It is also very difficult for people to judge just how much they can drink without exceeding those blood alcohol levels. Many people think it would be better to take the Russian approach, and say that anyone who needs to drive should not drink at all.

A 'safe' limit?

'What's a safe limit? It's only after a tragedy like this that people find their own limit – regardless of the law. Nobody who has had a drink should drive a car.'

Peter Harrison, after three members of his family were killed by a drunk driver, *Guardian*, 10 March 1998

The Russian police will prosecute anyone caught driving under the influence of alcohol. The amount you have drunk makes no difference.

Restricting drinking and driving works

'Put simply, lowering the drink-drive limit will save hundreds of lives.'

Eric Appleby, UK Alcohol Concern

- In 1994, drinking caused 510 deaths on the roads in the UK, compared with 1,650 in 1979.
- In 1995, 17,126 people were killed in crashes involving alcohol in the USA, compared with 22,715 in 1985.

Easier said than done

Everyone would like to see a reduction in the number of road accidents caused by alcohol. There is evidence to show that restrictions on drinking and driving, combined with public education about the dangers, do work, but those involved in the alcohol industry are opposed to further restraints on drinking. In the UK, alcohol restrictions for driving have changed little over the last 30 years. A recent government proposal to reduce the limit to one pint of beer (or the equivalent) put country pub owners up in arms. They argued that no one will bother driving out to their pubs if they can only have one drink when they get there. Many were worried that they would go out of business.

Country pub owners feel that stricter drinking and driving laws could put them out of business.

In the USA, pressure groups like Mothers Against Drunk Driving (MADD) have come up against much the same problem. MADD found that nearly 4,000 people were killed in alcohol-related accidents in 1995 by drivers under the legal alcohol limit. But its campaign to get the limit lowered has met considerable opposition from the American Beverage Institute, which represents some of America's most popular restaurant chains. Whose interests do you think are the most important?

The right to drink

Making whisky is a way of life for Arran islanders in Scotland.

Alcohol is not just a financial or business issue. Its long history in human society means it has become an important element in the culture of many countries. Not only is drink a familiar part of our social lives and rituals, it forms part of the national identity that gives each country its particular character. For example, where would Scotland be without its whiskies, France without its fine wines, Germany without its beer, or Russia without its vodka? It is hard to imagine what a teetotal world would be like.

Wine drinking in particular has an enormous cultural tradition. Knowing your fine wines is seen as a sign of education and sophisticated tastes. Across many countries, a whole culture has developed around wine drinking. There are clubs and societies that encourage people to try different wines. Some people spend their whole lives developing their knowledge of wine and its vintages.

47

Case study

After many years of study, Bob has become an expert on wines from all over the world.

Bob, an Australian car mechanic, has been studying wine for the last 15 years. He belongs to a wine appreciation society in Sydney, which meets twice a month to try different wines and learn about their production. Bob spends much of his spare income buying-in wines from around the world. He has collected over 1,000 bottles which he keeps cool in his cellar.

Every weekend, Bob tries a new wine and makes notes about its scent, flavour and vintage. Bob's friends consider him to be the local expert. Whenever they cook a special meal, they always ring him for advice on the right wine to go with the food. Bob says wine is the love of his life. His wife agrees, but luckily she is almost as fond of a good bottle of Australian Chardonnay as he is.

Appreciating fine wine is considered a social asset in many countries.

Whose life is it anyway?

Many people argue that drinking is a matter of individual choice. We should all of us, they say, have the right to choose what, when and how much we drink. Although they acknowledge that alcohol can cause problems, they do not see why its abuse by a minority should mean drink is banned for everyone.

Restrictions related to alcohol have been (and often still are) seen as an attack on people's freedom. The introduction of breathalysers in the UK in 1967 was extremely unpopular, because people felt they had a right to decide whether they were fit or unfit to drive and resented what they saw as an infringement of their liberty by a 'nanny' state.

Unfair to the majority

'Alcohol beverages are not responsible for the increase in health-care costs, and responsible drinkers should not be forced to pay a disproportionate share of the taxes needed to fund reform.'

Stephen K. Lambright, vice president and group executive, Anheuser-Busch, Budweiser brewery, USA

Should this man have the right to refuse the breathalyser?

Talking point

'I remember Breathalyser Day in 1967 and the vitriolic attacks on Castle the Killjoy, the enemy of freedom, the scourge of the moderate drinking classes.'

Neil Kinnock, European Commissioner for Transport, talking about the introduction by Barbara Castle, then Minister of Transport, of the first breath tests for drinking

Some people argue that the government has no right to interfere with how much people choose to drink. Those supporting alcohol restriction in areas like driving, however, believe that such laws protect the innocent. Who do you think is right?

Denying the benefits

Those who support the right to drink also argue that any complete ban would not only be a violation of people's right of free choice, but would also deny people access to the benefits of alcohol. Even today, alcohol is still used for medicinal and therapeutic purposes.

Alcohol is used in the treatment of patients as a source of easily absorbed calories when the digestive system needs to be bypassed. Alcoholic drinks are sometimes used to treat sleeplessness, and are occasionally used as a supplement in special diets. An alcohol ban would mean that a prescription would be needed to obtain alcohol for any of these uses.

A special case

Should we make a special case for alcohol? Those who support the decriminalization of cannabis argue that it is much safer and does far less damage than alcohol. In some Muslim countries the use of cannabis is widely tolerated whereas alcohol has been banned for the last 1,000 years.

But those who oppose cannabis say that just because alcohol is legal does not mean it is right to legalize another potentially harmful drug. Do you think alcohol is safe enough to be legal? If it was discovered today, do you think we would legalize it? Do we all have the right to drink?

If alcohol were invented tomorrow it would probably be classed as a dangerous substance and made illegal.

The human cost of alcohol

Alcohol can ruin people's lives. When people drink too much, it can have a devastating effect on them, their families and society as a whole. While drinking in moderation does not usually cause problems, a reliance on alcohol can have negative consequences, particularly for those who are emotionally or psychologically vulnerable.

Many down-and-outs drink to blot out the hopelessness of their situation. In time, having alcohol to drink becomes more important than eating properly.

Habit-forming

What can start as a couple of drinks to relax after a hard day's work can develop into a difficult habit to break. You might start drinking at lunchtime, as well as after work, and gradually come to depend on alcohol just to get through the day. In the worst stages of alcoholism, alcoholics no longer care about trying to lead a normal life, and drink during most of their waking hours.

Temporary suicide

'Drunkenness is temporary suicide: the happiness that it brings is merely negative, a momentary cessation of unhappiness.'

Bertrand Russell, philosopher

'One of the disadvantages of wine is that it makes a man mistake words for thoughts.'

Samuel Johnson, English writer

Although some people are more vulnerable than others, alcoholism can affect anyone, including the elderly and people you would not expect to drink to excess. It can make it impossible for people to hold down their job or have a decent social or family life. It can lead to a range of unpleasant physical symptoms and longer-term health problems. It can also cost a great deal of money.

Not just your problem

Unfortunately, alcoholism is not just a problem for the person who drinks. It affects the lives of everyone around them, particularly members of their family.

Alcoholism often has innocent beginnings, such as drinking with colleagues at lunchtimes or after work.

Talking point

'Whose is the misery? Whose the remorse? Whose are the quarrels and the anxiety? Who gets the bruises without knowing why? Whose eyes are bloodshot? Those who linger late over their wine, those who are always trying some new spiced liquor.'

Proverbs, 23: 29

Alcohol seems to have so many drawbacks. Why do you think it is still so popular?

Relationships often break up as a result of one partner's excessive drinking. Alcoholism can make people difficult to live with and often brings the added strain of money problems. It also increases the risk of domestic violence: many women who have been battered say that their partners only hit them when they have been drinking.

Having an alcoholic parent can also be very damaging for children. Imagine what it is like to fear violence or abuse when a parent has been drinking. They may be too drunk or hung over to look after you properly, and may spend so much on drink that they cannot afford the food and other things you need. Even if you are well looked after, it can be very hard to cope with and understand the mood swings that often accompany a parent's heavy drinking.

Some children find it very hard to cope with the extreme mood swings of an alcoholic parent.

Misery beyond measure

'Most parents can enjoy a drink or two without any problem. But no sensible person would drive a car or operate machinery while over the limit. Nor would a responsible carer be found drunk in charge of a child.'

Jim Harding, director of the National Society for the Prevention of Cruelty to Children, UK,
30 December 1997

'The human cost of this carnage which shatters lives and plunges entire families into misery is obvious and beyond measure.'

Neil Kinnock, European Transport Commissioner, 9 April 1997

Case study

Vicky lived with her mother and two brothers, Jason, 7, and Darren, 4. Their father left when Vicky was young. Their mother had been suffering from a drinking problem for 10 years. When she was sober she was a good parent, who loved cooking for her children and taking them to the park. But when she was drunk she could not feed them, take them to school or make sure they were safe around the house.

Vicky doesn't drink, but alcohol helped to ruin her childhood.

With her children taken away, Vicky's mother has only the bottle for comfort.

As a result, Vicky and her brothers often missed school for long periods and found it difficult to make friends. Vicky suffered from painful stomach aches for several days at a time and often felt that somehow she was to blame for her mother's drinking. Darren wet the bed nearly every night.

Friends and neighbours tried to help look after them, but got tired of their mother's promises to stop drinking. Eventually all three children were taken into care, when social services decided their mother was not fit to look after them. They are currently in a foster home, while a suitable family is sought to adopt them.

The wider picture

Alcoholism is not just a problem for individuals and their families. Whole communities can suffer because of alcohol dependency, especially where drink did not previously play an important part in their culture.

For these Australian Aborigines, alcohol is both a curse and a comfort.

In the USA, for example, alcoholism is a very big problem among many Native American communities. The breakdown of traditional social structures and Native American culture, the poor living conditions and even poorer prospects for work tempt many into using alcohol as an escape. But drinking too much can lead in turn to an increased sense of hopelessness and further disintegration of the community. Suicide rates are often high.

In Australia and New Zealand, similar problems with alcohol have developed within the Aboriginal and Maori communities.

Alcohol-related crime

Violence and assaults are common around bars, especially on Friday or Saturday nights when people may have been drinking heavily for several hours. We talk about people being 'under the influence' of alcohol, because it makes them more aggressive and lowers the inhibitions that would usually prevent them from acting rashly.

Alcohol also affects people's ability to think sensibly about the consequences of their actions. This makes it more likely that someone will overreact in a situation, or will take a risk they would normally avoid. When people have been in a fight or broken the law, they often blame it on being drunk.

Under the influence

UK research shows that offenders are intoxicated in:

- 30 per cent of sexual offences
- 33 per cent of burglaries
- 40 per cent of crimes of domestic violence
- 50 per cent of crimes in the street
- 85 per cent of crimes in pubs or clubs

Either offender or victim have been drinking in:

- 65 per cent of murders
- 75 per cent of stabbings

UK Alcohol Concern

Alcohol can heighten aggression, leading to crime and disorder.

Seeking help

Kicking the habit can be very tough. People who have a drinking problem say it stays with you for life. Even if they haven't touched a drink for 20 years, ex-alcoholics cannot afford to consider themselves cured. Most recovering alcoholics always have to resist the urge to drink, and they avoid alcohol for the rest of their lives.

Fortunately, many clinics and organizations exist to help people cope. Alcoholics Anonymous (AA), for example, began in 1935 when two alcoholics met up regularly to help each other stop drinking. AA is now a worldwide organization with branches in over 130 countries, 87,000 local groups and over 2 million members. Its 12-step recovery programme has helped many people conquer their drinking problems. Related groups such as Al-Anon and Alateen have evolved to support the families of alcoholics and teenage children affected by alcoholism.

Professional counselling can help alcoholics towards recovery.

Moderate drinking can form part of a healthy and active social life, and need not escalate into alcoholism.

Most countries have clinics where people can stay and tackle their drink problem. Treatment normally involves a ban on drinking and the opportunity for patients to discuss their alcoholism, with counsellors and other alcoholics. For those who are heavily dependent on alcohol, medically supervised detoxification programmes and drug therapies can help to break their dependency. It can be a long and difficult process, but alcoholism can be beaten.

Not all those seeking help are alcoholics, but they are misusing alcohol and might be in danger of alcoholism if their drinking is not controlled and reduced. Specialists supporting young people who are misusing alcohol work with them to reduce the number of units consumed each week to a more healthy level. In such cases, people are being positively encouraged towards moderate drinking which will present no danger to their health or that of those around them.

Glossary

Alcohol A colourless liquid produced from the fermentation of carbohydrates and yeast. Also known as ethanol or ethyl alcohol, it is the active ingredient of alcoholic drinks such as wine, beer and spirits.

Alcoholism Also known as alcohol dependence. An illness characterized by habitual, compulsive, long-term heavy consumption of alcohol. Sufferers develop withdrawal symptoms if they suddenly stop drinking.

Alcopops Alcoholic versions of soft drinks such as lemonade.

Blood alcohol A measure of how much alcohol is in a person's blood.

Breathalyser Instrument for measuring alcohol levels in the breath exhaled into it.

Brewing The method used to make beer, where grain is boiled in water which is then fermented.

Cirrhosis Damage caused to the liver due to excessive alcohol consumption.

Claret The English name for red wine from the Bordeaux region of France.

Cognac High-quality brandy. Under French law the name Cognac can only be given to brandy produced in a certain area around the town of Cognac, in the Charente region of western France.

Depressant A substance that lowers the physiological processes in the body such as the metabolic rate, heart rate etc., resulting in a 'slowing down' effect which influences reaction times and can cause people to feel emotionally lower.

Distillation A method of producing stronger drinks by heating and concentrating the alcohol.

Drunk When someone has consumed a lot of alcohol and is heavily affected by it.

Fermentation A process by which sugar is turned to alcohol by natural organisms like yeast.

Foetal alcohol syndrome A rare condition resulting from high alcohol consumption by a mother during pregnancy. Affected babies may be mentally handicapped and suffer from physical abnormalities.

Hangover Severe headache and feeling of sickness caused by consuming too much alcohol.

Hepatitis Inflammation of the liver.

Liqueur A sweet, flavoured alcoholic spirit.

Prohibition The forbidding by law of the manufacture and sale of alcoholic drinks.

Spirit The strongest type of alcoholic drink, such as vodka, whisky and brandy. Spirits are produced by the method of distillation.

Stout A type of beer made from roasted barley and malts.

Teetotal Abstinent from all alcohol.

Vintage The year's grape harvest. The term is also used to describe the date and character of the wine produced from that harvest.

Books to read

Alcohol (Preteen Pressures Series) by Pavla and Paula McGuire (Raintree/Steck Vaughn, 1998)

Alcohol Abuse: How to Help a Loved One by Pippa Sales (Disa, 1994)

Alcohol is a Drug Too: What Happens to Kids When We're Afraid to Say No by David J. Wilmes (Johnson Institute, 1993)

'Alcohol: Teenage Drinking' in *The Encyclopedia of Psychoactive Drugs* by Alan R. Lang (Chelsea House, 1992)

Alcohol and You (Impact Series) by Jane Claypool (Franklin Watts, 1997)

Drinking: A Risky Business by Laurence Pringle (William Morrow, 1997)

Let's Talk About Alcohol Abuse (The Let's Talk Library) by Marianne Johnston (Rosen Publishing Group, 1997)

Living with a Parent Who Drinks Too Much, by Judith S. Seixas (William Morrow, 1983)

'Prohibition: America Makes Alcohol Illegal' in *Spotlight on American History* by Daniel Cohen (Millbrook, 1995)

Saying No Is Not Enough: What to Say and How to Listen to Your Kids About Alcohol, Tobacco and Other Drugs – A Positive Prevention Guide for Parents by Robert Schwebel and Benjamin Spock (Newmarket, 1998)

Straight Talk About Drugs and Alcohol (Straight Talk Series) by Elizabeth A. Ryan (Facts on File, 1995)

A Young Person's Guide to the Twelve Steps by Stephen Roos and Margaret O' Hyde (Hazelden, 1988)

Sources

Buzz: The Science and Lore of Alcohol and Caffeine by Stephen Braun (Penguin USA, 1997)

The Mother's Survival Guide to Recovery: All About Alcohol, Drugs and Babies by Laurie L. Tanner (New Harbinger, 1996)

Websites

http://www.alcoholconcern.org.uk
Home page for the UK organization Alcohol Concern.

http://www.ncadd.org
American site with links to many other related sites.

http://www.nzdf.org
The New Zealand Drugs Foundation home page.

http://www.eurocare.org
A website with Europe-wide news, information and discussions about alcohol.

Useful addresses

UK
Alcohol Concern
Waterbridge House
32–36 Loman Street
London SE1 0EE
Tel: 0171 928 7377

Alcoholics Anonymous
PO Box 1
Stonebow House
Stonebow
York YO1 2NJ
Tel: 01904 644026 (local helpline
numbers are available in all regional
telephone directories)

Drinkline Youth
(For confidential advice about drinking)
Tel: 0345 320202

Health Education Authority
Trevelyan House
30 Great Peter Street
London SW1P 2HW
(A selection of leaflets about drugs and alcohol
are available for parents and young people)

TACADE (Teachers' Advisory Council on
Alcohol and Drug Education)
1 Hulme Place
The Crescent
Salford
Greater Manchester M5 4QA
Tel: 0161 745 8925

USA
Alcoholics Anonymous
PO Box 459
Grand Central Station
New York, 10163
Tel: (212) 870 3400

NCADD (National Council on Alcoholism
and Drug Dependence)
12 West 21 Street
New York 10010
Tel: (212) 206 6770

Europe
Eurocare
(Advocacy for the prevention of
alcohol-related harm in Europe)
1 The Quay
St. Ives Street
Cambridgeshire PE17 4AR
UK
Tel: 01480 466766

Australia
Alcohol and Drug Foundation
19–23 Townshend Street
Phillip
Canberra

Index

THE LEARNING CENTRE
HAMMERSMITH AND WEST
LONDON COLLEGE
GLIDDON ROAD
LONDON W14 9BL

0181 741 1688